MW01377781

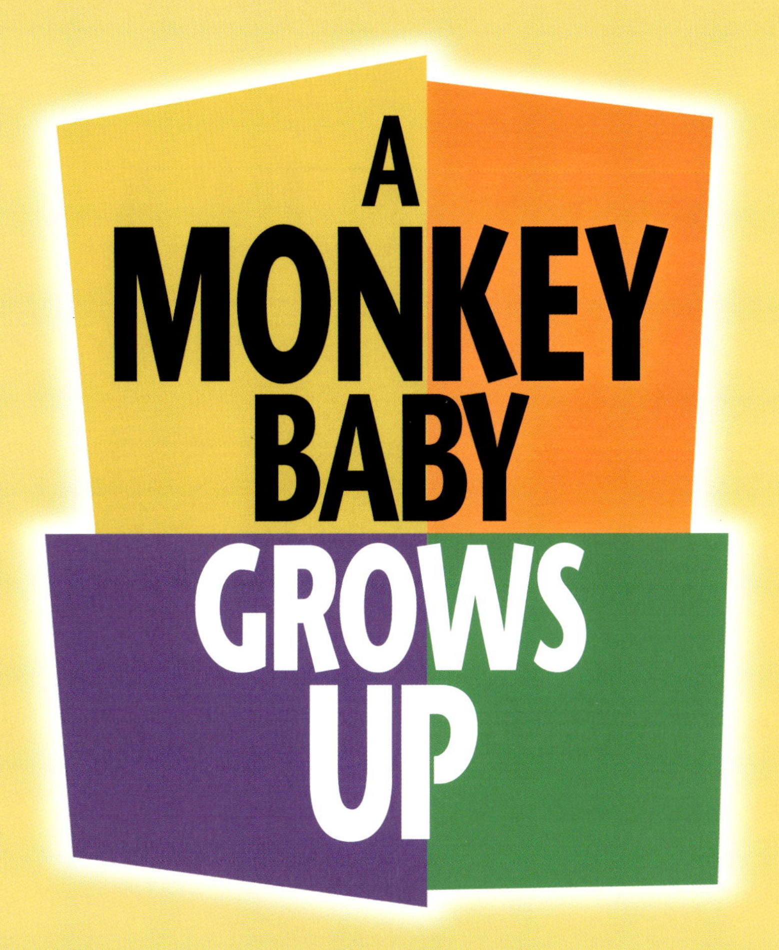

A MONKEY BABY GROWS UP

by Joan Hewett
photographs by Richard Hewett

Carolrhoda Books, Inc./Minneapolis

Born in a Zoo

Gabbie is a monkey baby.
She was born during the night.
Now the sun is up.

The newborn baby is alert.
She reaches out her arm.
She touches her mother's face.

Gabbie's father sees her.
He is excited.
"Whoo, whoo, whoo," he whoops.

Gabbie is hungry.
She drinks her mother's milk.

Gabbie's mother is hungry also.
She gets up.
It's time for breakfast.

Monkey babies have strong fingers and toes.
Gabbie grasps her mother's fur.
Then riding upside down,
she clings to her mother's belly.

A keeper brings out food.
There are vegetables, fruits, and nuts.

Gabbie's mother eats a banana.
Gabbie hangs on.

Breakfast is over.
Gabbie snuggles
in her mother's lap.
She feels safe.

Gabbie looks around.

Another monkey is watching her.

Gabbie sees other monkeys too.
The monkeys take care of each other.
One monkey grooms another.
The monkeys are like a family.

Gabbie just wants to be
with her mother.

READY TO EXPLORE

Gabbie is 2 months old.
She is curious.
She makes a soft chattering sound.

Gabbie is calling her mother.
Her mother comes.
Gabbie clings to her mother.
Off they go.

As Gabbie grows, her muscles get stronger.
She climbs on a tree branch.
She rolls over and plays.
She doesn't fall off.

The monkey baby is 4 months old.
She looks more and more like her mother.

Gabbie
is old enough
to eat solid food.
Her mother's
orange looks good.

It smells good.

How does it taste?
GOOD!

Gabbie waits for the keeper.
When she arrives, Gabbie greets her.

The keeper gives Gabbie a juicy treat.

Gabbie has plenty to eat.
She no longer drinks her mother's milk.

Still, Gabbie plays with her mother.
And her mother plays with Gabbie.

Gabbie Loves to Play

Gabbie is 6 months old.
She likes to explore.
She plays with other monkeys now.

Her father comes over.

Gabbie gives him a gentle pat.

Gabbie sees a pole.
Could it be a toy?

She pushes one end into the ground.
She grabs the other end and jumps.

How do you hold a scratchy stick?
Gabbie learns as she plays.
Playing makes her muscles stronger.

The monkey baby
is not grown up yet.
But she is on her way.

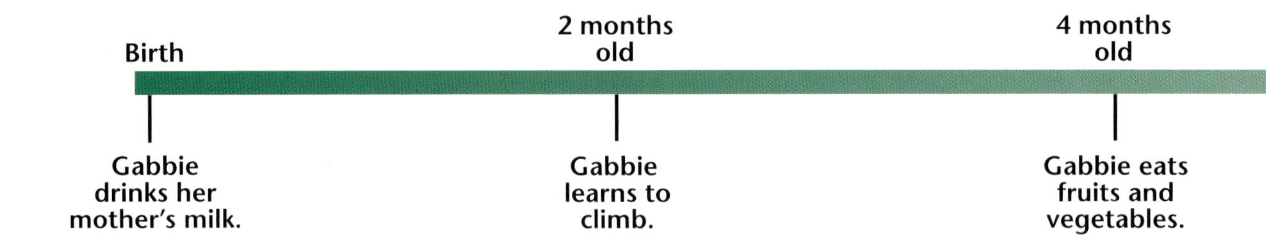
Birth
2 months old
4 months old
Gabbie drinks her mother's milk.
Gabbie learns to climb.
Gabbie eats fruits and vegetables.

More about Monkeys and the Red-Crowned Mangabey

Monkeys belong to a group of animals called primates. Chimpanzees, orangutans, and gorillas are primates. People are primates, too. Like most other primates, monkeys have five fingers on each hand and five toes on each foot. With the help of its thumb, a monkey can grasp branches and pick up and hold food.

The red-crowned mangabey is an African monkey. It is also known as the red-capped mangabey and white-collared mangabey. Red-crowned mangabeys range across hot humid rain forests and swamplands, from Nigeria to the Zaire River. The monkeys scamper up trees and vines. They leap through the air with dazzling ease. Still, red-crowned mangabeys spend most of their lives on the ground.

If you were walking through a rain forest in Nigeria, you would probably hear red-crowned mangabeys before seeing them. Their whooping, barking, grunts, and chatter would lead you to them. These monkeys live in groups. A group of mangabeys moves through the rain forest in search of food and water. The monkeys call to each other to keep in touch. If one monkey spies a crouching leopard, its loud cries will alert the others.

The rain forests where the red-crowned mangabeys live are shrinking. Trees are being cut down for lumber. Houses are going up. Swampland is being drained. Farmers are planting crops. The charming, bright mangabey is in danger of extinction.

More about Zoos

Most of us will never see a monkey in its natural habitat. But you can see monkeys at a zoo. Spend some time there. These lively animals are fascinating to watch.

Gabbie, the red-crowned mangabey in this book, lives at the Los Angeles Zoo in California. Other zoos have red-crowned mangabeys too. You can see them at Binder Park in Michigan, the Brookfield Zoo in Illinois, the Denver Zoo in Colorado, the Houston Zoo in Texas, the Kansas City Zoo in Missouri, and the Philadelphia Zoo in Pennsylvania.

INDEX

For our grandsons, Orson Ridgely, Jesse Angelo, and Nathan Morris

This book is available in two editions:
Library binding by Carolrhoda Books, Inc., a division of Lerner Publishing Group
Soft cover by First Avenue Editions, an imprint of Lerner Publishing Group,
241 First Avenue North
Minneapolis, MN 55401 U.S.A.

Website address: www.lernerbooks.com

Library of Congress Cataloging-in-Publication Data

Hewett, Joan.
A monkey baby grows up / by Joan Hewett ; photographs by Richard Hewett.
p. cm. — (Baby animals)
Summary: Describes the development of Gabbie, a monkey living in a zoo, from birth to age six months.
ISBN: 1–57505–199–0 (lib. bdg. : alk. paper)
ISBN: 1–57505–632–1 (pbk. : alk. paper)
1. Monkeys—Infancy—Juvenile literature. [1. Monkeys. 2. Animals—Infancy. 3. Zoo animals.] I. Hewett, Richard, ill. II. Title. III. Series.
QL737.P9H45 2004
599.8'139—dc22
2003011749

Manufactured in the United States of America
1 2 3 4 5 6 – DP – 09 08 07 06 05 04